Management Division 2020

Held at the 2020 AIChE Annual Meeting

Online
16 – 20 November 2020

ISBN: 978-1-7138-2300-1

Printed from e-media with permission by:

Curran Associates, Inc.
57 Morehouse Lane
Red Hook, NY 12571

Some format issues inherent in the e-media version may also appear in this print version.

Printed with permission by Curran Associates, Inc. (2021)

For permission requests, please contact AIChE
at the address below.

AIChE
120 Wall Street, FL 23
New York, NY 10005-4020

Phone: (800) 242-4363
Fax: (203) 775-5177

www.aiche.org

Additional copies of this publication are available from:

Curran Associates, Inc.
57 Morehouse Lane
Red Hook, NY 12571 USA
Phone: 845-758-0400
Fax: 845-758-2633
Email: curran@proceedings.com
Web: www.proceedings.com

TABLE OF CONTENTS

(92B) WHAT I LEARNED IN MY INDUSTRIAL POSTDOC: PROJECT MANAGEMENT SKILLS FOR ACADEMIA 1
Julie Renner

(92C) MANAGING AN INTERDISCIPLINARY TEAM OF UNDERGRADUATE AND GRADUATE RESEARCHERS 2
Shannon Servoss

(92D) LEADING TECHNICAL TEAMS FOR COMPLEX ENERGY AND SUSTAINABILITY PROBLEMS 3
Paul Dauenhauer

(92E) CREATING INDEPENDENT RESEARCHERS 4
Sarah L. Perry, Peng Bai

(198A) LEADING AND MANAGING INDUSTRY-ACADEMIA PARTNERSHIPS AND COLLABORATIONS PANEL DISCUSSION 5
Aditi Khadilkar

(389A) USE OF AGILE AND SCRUM IN PRODUCT DEVELOPMENT 7
Kathleen Dennison, Simon Banks, Jonathan Galownia

(389B) CROSS-FUNCTIONAL SCRUM 8
Austin S. Lin

(389C) ESSENTIAL FUNDAMENTAL PROBLEM SOLVING TOOLS 9
Zdravko Stefanov

(389D) LEAD IT! TOOLS AND STRATEGIES FOR EFFECTIVE PROJECT MANAGEMENT AND LEADERSHIP 10
Eldad Herceg

(537A) DEPOSITION OF OXIDE COATINGS USING NON-THERMAL ATMOSPHERIC PRESSURE PLASMAS 11
David Barlaz, Dhruval Patel, Zachary Jeckell, Daniel Krogstad, Brian Jurczyk, David Ruzic

(537B) TREATMENT AND EXTRACTION OF COPPER FROM ELECTRONIC WASTE VIA INDUCED MORPHOLOGICAL CHANGES UTILIZING SUPERCRITICAL CO2 12
Emily Hsu, Christopher Durning, Alan West, Ah-Hyung Alissa Park

(537C) ANNEALING CARBON BY PULSED LASER LIGHT 13
Randy Vander Wal, Akshay Gharpure

(537D) AN ALL-IN-ONE APPROACH FOR TRAINING DEEP LEARNING-BASED CONTROL LAWS 15
Yankai Cao

(537E) CLOUD-BASED CONTROL OF A ROBOTIC MANUFACTURING PROCESS FOR PERSONALIZED MEDICINES 18
Alice Melocchi, Zack Bright, Federico Parietti

(537G) EMERGING COGNITIVE ENGINEERING APPROACHES TO ENHANCE CONTROL ROOM OPERATOR CAPABILITIES IN CHEMICAL INDUSTRIES 20
Mahindra Choudhary, Md. Umair Iqbal, Babji Srinivasan, Rajagopalan Srinivasan

(284B) ENERGY LEADERSHIP TOWARD A SUSTAINABLE ENERGY FUTURE 23
Scott Klara

(284C) LEADERSHIP IN ENERGY DEVELOPMENT 24
Charles McConnell

(284D) MANAGING PUBLIC SECTOR TECHNOLOGY DEVELOPMENT IN A DYNAMIC ENERGY ENVIRONMENT 25
Jared Ciferno

(284E) LEADING THE NATION'S NUCLEAR CLEANUP AND PROVIDING SUSTAINABILITY 26
Kenneth Rueter

(141B) THE CHALLENGE FOR LEADERS TODAY 27
Paulette Clancy

(141C) BUILDING AND MAINTAINING A POSITIVE CULTURE IN AN ACADEMIC DEPARTMENT 28
David S. Sholl

(207A) INTRODUCTIONS OF PANELISTS 29
Meredith Sellers, Felicia Etzkorn, Motunrayo Kemiki, Tricia Berry, Jessie Dearo, Ekin Gozen

(229B) THE DIGITAL TRANSFORMATION OF THE CHEMICALS INDUSTRY - AND HOW TO PREPARE 31
Michael Holman

(229C) DRIVING THE VALUE FROM DIGITALIZATION 32
Billy Bardin

(229D) COMBINING AI WITH CHEMICAL ENGINEERING FIRST PRINCIPLES WILL TAKE PROCESS PERFORMANCE, SAFETY AND SUSTAINABILITY TO NEW LEVELS 33
Ron Beck

(229E) DESIGNING MATERIALS AND PROCESSES IN THE DIGITAL WORLD 34
Rob Kieschke

Author Index

601998 What I Learned in My Industrial Postdoc: Project Management Skills for Academia

Abstract:

For many, a postdoctoral position is the next major career step in their career path after obtaining a Ph.D. A large majority of postdocs, 89%, will be in academic labs. The rest receive their training in industry, national laboratories or other government entities. While industrial postdocs are relatively rare, the experience offers a variety of benefits, including the opportunity to participate in project and grant management. Here, we specifically discuss the benefits of industrial postdocs in gaining project management experience. The hands-on training can provide strategies for assembling and managing a team, project goals, budgets, and collaborations. How those skills can be translated to successful academic grant management will be highlighted.

613042 Managing an Interdisciplinary Team of Undergraduate and Graduate Researchers

Abstract:

The success of grants is dependent on the organization of research efforts. This is especially important when the researchers have different areas of expertise, as well as different levels of experience. In this presentation I will discuss managing an interdisciplinary team of researchers from both the undergraduate and graduate level.

606309 Leading Technical Teams for Complex Energy and Sustainability Problems

Abstract:

The sustainability challenges of the 21st century are sufficiently complex to require a team approach of diverse skills and backgrounds. The development of carbon-free energy, a circular economy of recyclable materials, and long-term sustainable transportation and clean water systems will only occur with transformational approaches which combine technical, social, and economic solutions. Leading technical teams to address these problems has its own unique challenges. Academics working on these chemical technologies need to communicate and work across skill areas including research subfields of simulation and computational chemistry, process systems and analysis, materials synthesis and characterization, and experimental chemistry and catalysis. Moreover, technologies emerging from these technical teams must integrate with social expectations and economic restrictions; the teams must both share and receive feedback on the viability of their ideas for both implementation with society and commercialization across many types, sizes, and structures of companies. In this presentation, best practices and experiences are shared on maximizing the success of academic teams taking on the most important challenges of the coming decades. Specific examples include an approach to selecting problems, organizing teams, data and idea sharing, and communicating long- and short-term strategy. Finally, a discussion is provided on strategies for pivoting to new solutions and goals as insights emerge and inform the problem.

608011 Creating Independent Researchers

Abstract:

In my experience, writing a grant is typically more about scientific questions and experimental procedures than it is about the details of actuating the project in the real world. However, once the grant is funded, the challenge shifts from one of scientific inquiry to project management. This raises the challenge of achieving a stated set of scientific and/or technical goals while working to educate and empower the individuals (particularly students) who are involved in the research. Managing a research group and mentoring students is often cited as one of the most challenging aspects for new PIs. I will discuss my strategies for creating a vibrant and collaborative research group that meets the individual members where they are and empowers them to take ownership over a research project and grow as an independent scientist.

612375 Leading and Managing Industry-Academia Partnerships and Collaborations Panel Discussion

Abstract:

Outline:

- Discussion on benefits resulting from and concerns hindering industry-academia collaborations
- New methods and long term changes needed to lead and manage industry- academia partnerships and collaborations
- Benefits to diverse sectors including R&D, manufacturing, economics, strategic planning etc. and increasing participation across all.
- Definition of research contracts, communication channels, international collaborations, project timelines and deliverables

Invited Panelists:

Barclay Satterfield: External Innovation Leader at Eastman Chemical Company who has experience in open innovation and industry/university research partnerships, polymer property research and environmental life cycle analysis. She has been involved in a $10M 6-yr research partnership between Eastman Chemicals and NC State University.

Kenneth Rueter: President and CEO - UCOR LLC, a past Management Award recipient has over forty years of experience. He has been involved in a highly successful collaboration with the University of Tennessee Nuclear Engineering Department. This collaboration has established a Nuclear Decommissioning minor at the University of Tennessee enabling students to meet the high demand for Nuclear Legacy Cleanup around the World.

Dr. Kiarash Vakhshouri: Young professional working with Google with a solid expertise in structural characterization of semiconductor materials, thin film coating and printing, rheology, lithography and designing and operating laboratory and pilot plant chemical processes. He has several well-cited publications and four US patents to his credit and has been engaged in a number of industry-academia collaborative research projects right since graduate school.

Dr. Phil Westmoreland: Has a PhD from MIT and has been a Professor of Chemical and Biomolecular Engineering with the North Carolina State University. He has over 35 years of academic research experience and has held a variety of Directorial positions during his illustrious career. He has also served as the AIChE President in 2013. His research interests include flame chemistry, theoretical and computational chemistry and polymer decomposition kinetics and thermochemistry. He has been instrumental to a number of industry-academia-government collaborative projects.

Dr. Quinta Warren:CEO at Energy Research Consulting, PE, PMP has experience in industry, academia as well as Government. She was working with ConocoPhillips, has been a director of professional education and also worked with the US DOE. She brings in a unique entrepreneurial skill-set into synergizing these areas.

Dr. Sarma Pisupati: Professor of Energy and Mineral Engineering/Chemical Engineering and Director of Center for Critical Minerals at Penn State. He has more than 35 years experience in academia and has prior industrial experience as well. He specializes in computational fluid dynamic modeling of combustors, coal/biomass gasification, and biomass based materials for emission reduction. He has driven several successful collaborative research projects and novel initiatives with the US DOE as well as industrial partners.

Samuel Tam:(Have reached out only this week and waiting to hear back) President and CEO of CAC Solutions LLC and has been with the US DOE for 9 years where he managed total annual R&D spending was over U.S. $100 MM. He also has experience as a China Chief representative while working with the US DOE. Over his forty years of R&D, engineering, business development, pilot plant design and operation, and commercialization of technologies related to climate change in US DOE and five other major US corporations, Samuel has gained immense experience on fostering collaborations.

606197 Use of Agile and Scrum in Product Development

Abstract:

Agile and Scrum have a long history of success in developing and deploying software products. We will describe our experience building an agile mindset and implementing agile frameworks, particularly Scrum, in product and technology development in a diversified manufacturing setting. Topics covered will include what we've learned about the preconditions and leading indicators of successfully using these approaches in a variety of businesses, lessons learned and adaptations made in using the Scrum framework, and some thoughts on measuring success. Case studies will be used to illustrate these points.

603607 Cross-Functional Scrum

Abstract:

You had the perfect Gantt chart and project schedule, but your project still got delivered behind schedule and above budget. Why?

Traditional project management tools are well intentioned, but oftentimes are weakened when the unexpected event arise, from changes in business strategy, fluctuations in the global economy, to more recently, health pandemics.

Scrum takes an action-oriented approach that adapts projects to their daily realities. This workshop will provide an overview of the scrum process (sometimes also called "agile" but as Jeff Sutherland, the method's founder would tell you, it's called "scrum" for a reason). We will then proceed into a hands-on workshop that illustrates how having a high powered cross-functional team further energizes the methodology with participants conducting an active scrum project in the session.

603926 Essential Fundamental Problem Solving Tools

Abstract:

The discipline of Fundamental Problem Solving is a very successful structured approach that can be applied to any problem that needs solving. As such, it attacks the most complicated problems in the company. To achieve success, it utilizes a wide range of tools, including project management tools, engineering tools, analytical tools, etc. In this session, we will cover the most widely used tools in the Fundamental Problem Solving discipline arsenal and provide examples of their application. This session will give a chance to young engineers to get acquainted with the Fundamental Problem Solving toolset and help them successfully solve problems in their plants.

603610 Lead It! Tools and Strategies for Effective Project Management and Leadership

Abstract:

This talk will focus on technical projects and how to manage and lead them effectively. Some of the most common tools and the strategies on how to deal with people issues will be presented. Emphasis will be given to dealing with individuals stalling the projects, commonly known as NoNos. Servant leadership and conflict triangle will be discussed as well as the importance of networking, recognition and project closing steps.

606167 Deposition of Oxide Coatings Using Non-Thermal Atmospheric Pressure Plasmas

Abstract:

Atmospheric pressure plasmas (APPs) have shown great promise in a variety of manufacturing applications. APPs offer unique opportunities for the industrial application of plasmas, due to the versatility of the technique, the ease of scalability, and the opportunity to perform the technique point of manufacturing. This work focuses on depositing zirconia-silica bilayer coatings as a replacement for hexavalent chromium-based coatings. A new non-thermal process to deposit zirconia was developed using APPs enabling processing on temperature sensitive substrates. Electrochemical corrosion performance will be compared to other passivation and barrier coatings. Additionally, we investigate APPs to enable lightweight manufacturing for the automotive industry by depositing a thin film of silica that acts as a water intrusion barrier on bonds joints between dissimilar materials. The adhesion tests performed after water soak meets the industry benchmarks. These processes aim to replace wet chemical processes often used by the automotive and aviation industries. The mechanical properties of the films are tested using lap shear, both before and after a water soak test, which successfully demonstrates the films capabilities as a water barrier. The chemical composition is confirmed using both XPS and RBS, and the films morphology is examined using cross sectional SEM.

608084 Treatment and Extraction of Copper from Electronic Waste Via Induced Morphological Changes Utilizing Supercritical CO_2

Abstract:

The management of electronic waste (e-waste) has become a significant challenge faced by humanity. Globally, the number of discarded computers, phones, and other appliances, often defined as Waste Electrical and Electronic Equipment (WEEE), doubled between 2009 and 2014 to 41.8 million tons per year. Currently, e-waste disposal factories tend to recycle the non-toxic, valuable components of e-waste (metals) and release the toxic components into the surroundings via pyrometallurgy (burning). Chemical treatment, known as hydrometallurgy, is also used and involves large amounts of corrosive acids. As the amount of e-waste grows, the availability of easily mineable elements, particularly metals, decreases, disturbing the overall material cycle. Thus, new, environmentally-friendly approaches to recover valuable components from e-waste are urgently needed. This study focuses on the use of supercritical CO_2 ($scCO_2$) and sulfuric acid, 2 M H_2SO_4, as a novel solvent system for the treatment of e-waste, specifically in the extraction of copper. Printed circuit board (PCB) was used as the e-waste of study and is generally composed of 40% metals, 30% plastics, and 30% refractory material. Through the synthesis of metal and polymer composites, which reduced the complexity of PCB, the effects of $scCO_2$ in the solvent system were investigated. The results showed that $scCO_2$ and acid induced a permanent morphological change in the polymer component, creating voids and causing delamination. This finding was translated to the actual waste PCB system and a novel two-step process was developed. This two-step process involved the pre-treatment of small pieces of waste PCB with $scCO_2$ and 2 M sulfuric acid at 120 °C and 148 atm for 30 minutes followed by leaching of the pre-treated PCB in a solvent containing 2 M H_2SO_4 and 0.2 M H_2O_2 at ambient conditions.Experimental results showed that 82% of the copper contained in the PCB was extracted in under 4 hours. The characterization of the PCB demonstrated that the pre-treatment with $scCO_2$ and acid induced the crystallization of the plastics (polymer component), creating pores and weakening the structure of the PCB, thus enhancing transport of the solvent to the buried metal interfaces. This novel process utilizing $scCO_2$ minimizes intensive physical processing (e.g. grinding of the PCB) and reduces acid usage during the extraction of Cu from e-waste, presenting a greener alternative to the current methods of recycling of metals from e-waste, which tend to be energy-intensive and hazardous.

596648 Annealing Carbon By Pulsed Laser Light

Abstract:

Laser processing of materials is not new, but, using lasers to anneal, bond and otherwise transform carbons is. Fundamental understanding of the dependence upon carbon structure, morphology and chemistry is critical to implementing this technology into manufacturing and processing applications. In this work a Q-switched Nd:YAG laser and a continuous wave CO_2 laser are used to anneal carbon materials. Lasers provide rapid heating and cooling with high temporal control. The extent of transformation is kinetically controlled by time above the threshold temperature for transformation. Enabling the production of carbon structures not possible via traditional furnace annealing. The potential technological importance of laser annealing carbon is demonstrated. Examples of material processing and synthesis not possible via traditional furnace annealing are provided. To resolve the nanostructure changes HRTEM is employed.

Perhaps the most over looked application of laser annealing carbon is the ability to do so continuously and rapidly. Graphitization furnaces operate as a batch processing system. Time from start to completion is on the order of a day due to the slow (~25 °C/min) heating rates, long hold times, and slow cooling of the heavily insulated furnace. Additionally, these furnaces require routine maintenance and replacement of expensive specialty graphite heating elements. CO_2 laser annealing provides equivalent material transformation on the order of seconds and modern CO_2 lasers are designed for years of maintenance free use [1]. Although laser annealing is limited to thinner material due to limited heating depth, materials can be annealed continuously and with potentially high throughput.

The material was heated to 2,600 °C in 1.4 ms Under the action CO_2 laser radiation carbon can be heated to 2,600 °C in 1.4 ms. After 20 s, the structure is equivalent to that obtained from furnace annealing at 2,600 °C for 1 hr. as has been shown elsewhere [2]. In contrast a Q-switched Nd:YAG laser with pulse width ~ 8 ns heats carbon materials to graphitization temperatures with a heating rate on the order of a few 10^{11} °C/s. The peak temperature is controlled by the laser pulse energy; typical temperatures range between 2,500 °C to the C_2 sublimation temperature of 4,184 °C [2,3]. The time at elevated temperature is limited (time above 2,000 °C is 1.5 μs). On this time scale, the long-range material motions are kinetically restricted. Consequently the Nd:YAG laser annealing trajectory deviates from traditional furnace pathways. The limited time at elevated temperature can be used for the purpose of surface modification. Surface modification via kinetically limited oxidation is one potential application. Another application is making connections between carbons without additional material. Laser sintering a mixture of carbon materials could result in a wide range of potential applications.

Heat-Treatment-Temperature (HTT) and resulting carbon structural transformation has been extensively studied. A detailed quantification of HTT and resulting material annealing was provided by Oberlin in 1984 [4]. The four stages outlined in Oberlin's HTT diagram (temperature dependent) are believed to be separated by "very rapid" transitions [8]. So rapid, in fact, that intermediate structures are not observable on furnace heating timescales (several minutes minimum) due to slow ramp rates and have thus been ignored. These intermediate structures remain unexplored and represent an entirely new class of carbon materials. It has been demonstrated that CO_2 laser heating results in equivalent end structure as compared to furnace annealing [1]. The high temporal control of

CO_2 laser annealing allows for exploration of such intermediate structures that can be captured by the rapid heating and cooling timescales. In summary lasers are poised to be instrumental in the advancement of carbon science and technology.

References

[1] J.P. Abrahamson, Pulsed laser annealing of carbon, Ph.D. Thesis, Penn State, 2017.

[2] J.P. Abrahamson, M. Singh, J.P. Mathews, R.L. Vander Wal, Pulsed laser annealing of carbon black, Carbon 124 (2017) 380–390. doi:10.1016/j.carbon.2017.08.080.

[3] H.A. Michelsen, C. Schulz, G.J. Smallwood, S. Will, Laser-induced incandescence: Particulate diagnostics for combustion, atmospheric, and industrial applications, Prog. Energy Combust. Sci. 51 (2015) 2–48. doi:10.1016/j.pecs.2015.07.001.

[4] A. Oberlin, Carbonization and graphitization, Carbon 22 (1984) 521–541. doi:10.1016/0008-6223(84)90086-1.

604429 An All-in-One Approach for Training Deep Learning-Based Control Laws

Abstract:

Model predictive control (MPC) provides a general framework for controlling nonlinear systems. However, the high online computational latency limits the scope of its application for systems with fast dynamics. One way to enable the real-time application of MPC is to approximate the implicit MPC control law with a neural network [1,2,3]. The hypothesis of this approach is that a deep learning neural network can sufficiently approximate the nonlinear behavior of MPC, but at a negligible online computational cost that is significantly better than that of a full-fledged MPC. For linear systems with quadratic cost function, a neural network with rectified linear units as activation functions can exactly represent piecewise affine optimal control laws [4].

An essential aspect of the above-mentioned approaches is the training of large dimensional neural networks. Many researchers follow the "optimize then train" protocol. The idea in these algorithms is to simulate the optimal controller, generate corresponding state and optimal control action data pairs and then use these data to train a neural network via supervised learning to approximate the original optimal controller. Unfortunately, this conventional "optimize then train" protocol is not guaranteed to work on nonlinear systems. For nonlinear systems, it is possible that multiple optimal control actions exit for the same initial state. In this case, the MPC control law is a set-valued function. The non-uniqueness of the state-action data pairs makes it impossible to learn the underlying MPC control law through this two-step approach, as illustrated in Figure 1. Even if the uniqueness of optimal control actions is guaranteed, the optimal control problem might have multiple local optimal control actions, which are typically identified using local solvers. Moreover, for general nonlinear systems, no neural network structure with finite neurons can guarantee an exact representation of the MPC control laws. The training error of the neural network can lead to sub-optimal control actions and state constraint violations. The resulting errors could potentially accumulate over time and eventually making the neural network deviate significantly from the optimal control.

We propose a new "optimize and train" method that combines the steps of data generation and neural network training into one single large-scale stochastic optimization problem. This approach directly optimizes the closed loop performance of the DNN controller over a number of possible initial states. Directly optimizing the control law instead of control actions avoids the difficulty caused by the nonuniqueness of optimal control actions, as shown in Figure 2. Moreover, all constraints are guaranteed to be satisfied for all initial state scenarios considered in the optimization, while this guarantee does not hold for the conventional "optimize then train" method. We also propose methods to ensure constraints satisfaction for all possible initial states. The critical challenge is the solution a large-scale nonlinear stochastic optimization problem. However, tremendous progress has been made recently in the field of stochastic optimization that allows us to design and implement efficient local and global algorithms [5,6]. Besides directly using stochastic optimization solvers, we also proposed a recurrent neural network formulation for nonlinear systems with input constraints and soft state constraints. This approach can take advantage of existing packages (e.g., TensorFlow) in the machine learning community and can potentially tackle extreme-scale problems efficiently with GPUs. The benefits of our all-in-one approach over the conventional ``optimize then train" protocol is illustrated through numerical results.

References:

[1] Parisini, T., & Zoppoli, R. (1995). A receding-horizon regulator for nonlinear systems and a neural approximation. Automatica, 31(10), 1443-1451.

[2] Csekő, L. H., Kvasnica, M., & Lantos, B. (2015). Explicit MPC-based RBF neural network controller design with discrete-time actual Kalman filter for semiactive suspension. IEEE Transactions on Control Systems Technology, 23(5), 1736-1753.

[3] Chen, Y., Shi, Y., & Zhang, B. (2018). Optimal control via neural networks: A convex approach. arXiv preprint arXiv:1805.11835.
[4]Karg, B., & Lucia, S. (2018). Efficient representation and approximation of model predictive control laws via deep learning. arXiv preprint arXiv:1806.10644.
[5] Cao, Y., Zavala, V. M., & D'Amato, F. (2018). Using stochastic programming and statistical extrapolation to mitigate long-term extreme loads in wind turbines. Applied energy, 230, 1230-1241.
[6] Cao, Y., & Zavala, V. M. (2019). A scalable global optimization algorithm for stochastic nonlinear programs. Journal of Global Optimization, 75(2), 393-416.

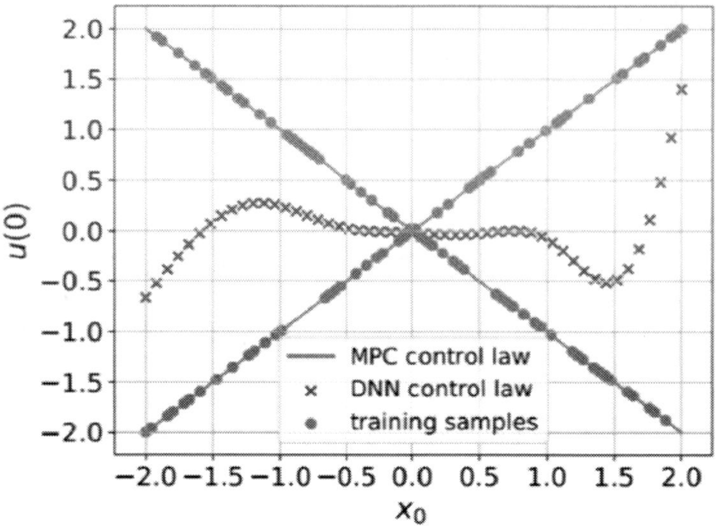

Fig. 1. DNN control law trained with "optimize then train" method

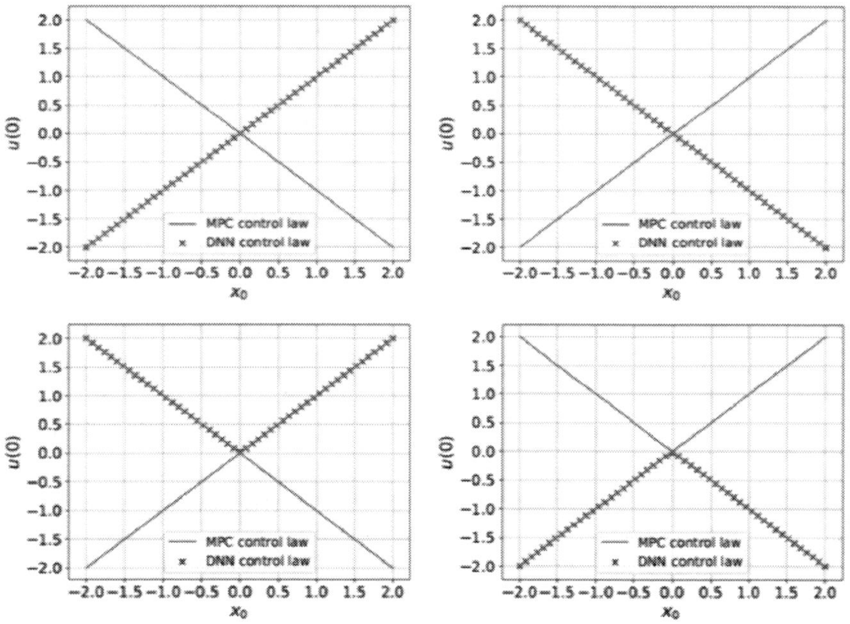

Fig. 2. DNN control law trained with "optimize and train" method

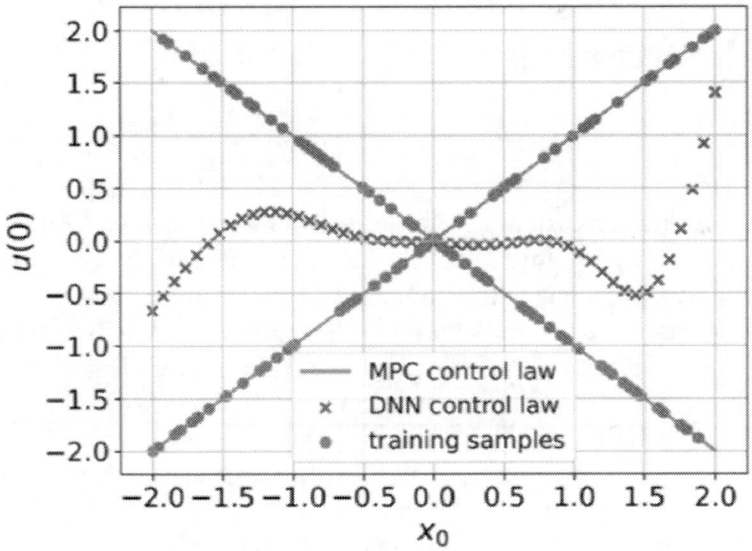

Fig. 1. DNN control law trained with "optimize then train" method

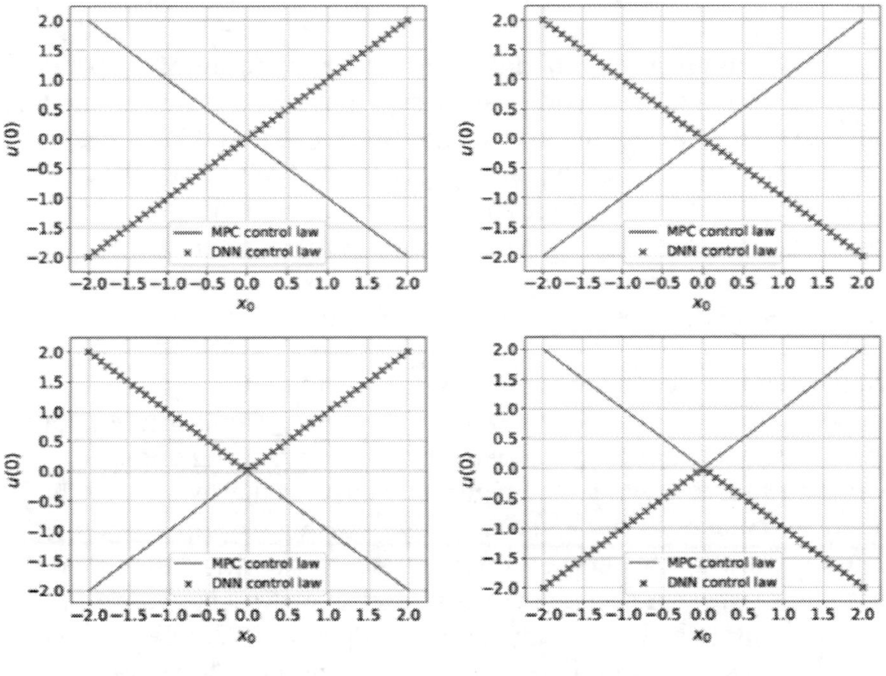

Fig. 2. DNN control law trained with "optimize and train" method

601990 Cloud-Based Control of a Robotic Manufacturing Process for Personalized Medicines

Abstract:

This abstract focuses on an automated production process enabling the manufacturing of personalized drug products. Particularly, we developed a system able to fulfill the needs of small-batch, high-mix, personalized, or even on-demand, production. Recently, the concept of precision medicine has received great attention. Finding ways to manufacture personalized therapies, in a consistent and controlled way is becoming the holy grail in the pharmaceutical field. However, existing manufacturing processes are not suitable for this mass-customization approach. In fact, while traditional systems can repeatably manufacture hundreds of thousands of identical dosage forms, having being designed for one specific product at a time, personalized therapies (*i.e.* extemporaneous formulations) are produced by technicians on a small scale only. Additionally, most of these production methods require labor to monitor, regulate and complete material handling for the equipment in use.

Based on these considerations, the availability of a fully automated architecture (*i.e.* hardware and software) capable of manufacturing personalized pharmaceutical products at scale would be of utmost importance. In this respect, we have developed a modular manufacturing system and a supporting cloud-based software architecture for its control. The modular manufacturing system, referred to as "robotic cluster", consists of *i)* different manufacturing modules, each of them performing specific fabrication steps, and *ii)* a transfer system that moves work-in-progress units (WIPUs) from module to module in a distributed and parallelized way. During the manufacturing process, WIPUs are shuttled into a given module, which performs a specific step of the process. Then, the WIPUs are moved into another module, and this process is repeated for all the following operations, including quality control. The sequence of manufacturing operations for a small batch of WIPUs is performed by moving it through the modules until all needed operations are completed. This sequence can be performed in a fully-planned, fixed-scheduled, a flexible-scheduled, or dynamic manner.

Each manufacturing module is connected to a dedicated control system and to other equipment *via* its control box, enabling the input of commands for the next fabrication steps while transmitting out the data pertaining to the work that has already been performed. All critical process parameters are monitored locally and in the cloud. The strategy pursued here not only ensures the security of the manufacturing system, but also increases its versatility and flexibility. It also makes this architecture suitable for scaling, by deploying many robotic clusters at the same time. Moreover, unlike flexible manufacturing systems described in literature, ours is comprised of a set of swappable units that are readily changed, updated, repaired and maintained. This way, complex operations are divided into simpler steps and for each of them there is a corresponding custom-built module. Therefore, the principal application of this technology is in mass customization, where products can be fully personalized to each order and large quantities of products can be fabricated in parallel to achieve the same scale and throughput of traditional serial manufacturing.

As a proof of concept, we evaluated the suitability of the discussed architecture for the manufacturing of multi-compartment capsules for personalized therapy. The capsules here proposed have a modular structure that consists of multiple inner compartments, which are independent of one another. Each compartment can be filled with different active ingredients and with a range of doses, as well as formulations, for any given drug. Multi-compartment capsular devices are especially advantageous for therapies involving multiple drugs, because they simplify the dosing schedule and have a positive potential impact on overall patient compliance.

Personalized filling of each compartment of the capsule shells was performed thanks to a custom-built robotic cluster, composed of a set of automated filling modules working simultaneously, in parallel. A robotic arm (designed to safely collaborate with human operators) moves the capsular devices form one filling station to the next, and quality control machines check the deposition of the desired amount of active ingredients after each step.

Within this proof of concept, we performed an optimization on the manufacturing steps as a fixed-scheduled sequence. We created a list of variable orders *i.e.* virtual patient prescriptions of varied precise dosages of multiple drugs, and input them into a schedule optimization algorithm. Said algorithm outputs the sequence of robot arm and module actions which maximizes throughput of the cluster. Generally, this algorithm maximizes the amount of work being done in parallel at once (many batches worked on in separate modules), and schedules the order in which the cluster outputs patient-scale batches to minimize the time taken to complete all batches. To ensure quality control we bound the time between manufacturing step and quality control: in-process controls should be measured more-or-less contemporaneously with the action. No WIPU can wait for longer than 10 min - say, sitting in a buffer or idle in a module - before being processed by the corresponding quality control module. For example, we bound the time between a filling module dosing powdered formulations and a weighing module recording the dosed weight. This scheduling optimization algorithm enables the parallelism of the robotic cluster. It is through this novel planning algorithm that we achieve the same scale and throughput of traditional serial manufacturing.

The automated deposition stage allows the rapid and accurate adjustment of the combination and dosage strength of the active ingredients. Cross-contamination is avoided by enclosing each filling station in a dedicated clean room. The capsule filling systems were validated with preliminarily filling trials, which used powders with different particle size distribution and bulk density. We considered 25 batches of 50 capsule each for each powder formulation. In all cases, the amount of active ingredient deposited by the robots had an average coefficient of weight variation (*i.e.* the ratio between the standard deviation and the average value of the weight of the deposited powder) below 3%. These results demonstrate the accuracy and the flexibility of the powder deposition systems. Additionally, we use spectroscopic techniques to check the weight uniformity and identity of the pharmaceutical formulations. Every measurement is collected and saved in the cloud based quality control system, allowing the creation, in real time, of a digital master batch record with all the information relevant to each manufacturing step (*e.g.* time, date, type of machine, weight of drug deposited), which could be accessed only by verified personnel and regulatory entities.

611934 Emerging Cognitive Engineering Approaches to Enhance Control Room Operator Capabilities in Chemical Industries

Abstract:

Control room operator's actions can have significant effect on the throughput, quality and safety of process plants (Baron, 1998). Typically, control room operators monitor the process via a Human Machine Interface (HMI) on which all the necessary information about the process state is presented. The complexity of operating modern chemical plants have increased significantly due to tight mass and energy integrations, deployment of sophisticated control automation strategies, and the need to operate the process in agile fashion in order to capture market and supply chain dynamics. However, there has been no concomitant increase in operations personnel's cognitive ability. Any mismatch between the operators' capability and the process control demands leads to high cognitive workload in human operators, often a precursor for poor performance which in-turn leads to sub-optimal operator actions or at worst accidents. This paper addresses the critical need to understand and enhance control room operator's cognitive capabilities.

Traditional operator performance assessment methods ignore cognitive aspects of performance. Operator's performance is assessed primarily in terms of number of successes and failures, response times, deviation of operator actions from predefined sequence (Lee et al., 2000), etc. However, in addition to overt performance (as done traditionally) it is imperative to consider covert aspects of performance i.e., cognitive workload for a holistic assessment of human performance. Various emerging cognitive engineering based sensors such as eye gaze tracking, Electroencephalography can reveal insights into the operators' cognitive state that determines these covert aspects. Specifically, in this paper we propose an Electroencephalography (EEG) based methodology to assess the operators' cognitive workload, especially during abnormal situation tasks where their active involvement is essential to safe process operations. The potential of EEG to understand cognitive workload for dynamic tasks which evolve with time has received some attention in aviation (Liang et al., 2018) and driving (Morales et al., (2017). However, there are hardly any studies relating to chemical control room operators.

We propose a single dry electrode EEG based methodology to dynamically assess the cognitive workload of operators while they perform various actions to control the process during abnormalities. The proposed methodology is illustrated on a simulated ethanol production plant testbed. A set of 10 different operators were considered in this work. Each operator's performance during the course of experiments involving monitoring and control of the process was studied. In addition to the process data and operator actions, we recorded the EEG data of the operators. Each experiment involved 6 tasks (comprising 5 different scenarios, out of which was repeated) of different abnormal situation. Each operator repeated the same experiment 7-8 times. Overall, 73 experiments were conducted by the 10 operators for a total of 438 tasks. Due to data quality issues, we evaluate results for 8 participants only.

We analyze the EEG signal in frequency domain by applying Fourier transform. Our results reveal that the power spectral density (PSD) of brain waves (4-30 Hz) is a biomarker that can identify the workload of operators. The PSD data of each operator was partitioned into two clustered using k-means clustering. Cluster 1 represents low power spectral density with low magnitude and Cluster 2 showed high power spectra density. We studied the workload variation with the help of cluster fraction over the course of the task as the operators' engagement with the process increased as well as with repetition. To evaluate the cluster variation within each task, a task is divided in three phases. In Phase 0 the operator monitors the process and ensures all the process variables are within the desired limits. During an abnormal condition an alarm is triggered, which leads to Phase 1 where the operator uses his mental model of the process to decide the control action(s) in order to clear the alarm. The portion of the task corresponding to all the control actions taken by the operator is called as Phase2. We observe distinct differences in the cluster fractions across the 3 phases. For a typical operator who successfully completes a task, during the monitoring Phase 0 we found the fraction of Cluster 1 to dominate (about 60%) indicating a correspondence between low workload during monitoring. During abnormality when alarms are triggered, in phase 1 the Cluster 1 fraction decreases (to about 40%) indicating that the operator uses his mental resources to decide the control action, i.e., increased workload. Further in phase 2, as the operator takes various control actions the Cluster 1 fraction further drops significantly (to about 0.20%) until the process becomes normal at which juncture it increases back to the 60% observed in Phase 0. The above patterns of change in the cluster pattern was observed for 81% (243 out of 300) of the tasks conducted by the 8 participants. This brings forth the capacity of the high accuracy of the EEG based cognitive sensor and the proposed machine learning algorithm to accurately estimate the workload of the control operator during normal and abnormal conditions. These distinct patterns were also highly correlated various process metrics (extent of variable deviation beyond alarm limits, time taken to clear alarm, number of standing alarms etc). In this paper, we will describe the proposed experimental and computational methodology and reported detailed results from our human subject experiments.

References:

Baron, RG. (1988). Human Factors in the Process Industries. Human Factors and Decision Making: Their Influence on Safety and Reliability, Symposium for the Safety and Reliability Society. ed. Sayers, BA. (pp.1-9)

Heard, J., Harriott, C. E., & Adams, J. A. (2018). A survey of workload assessment algorithms. IEEE Transactions on Human-Machine Systems, (99), 1-18.

Srinivasan, R., Srinivasan, B., Iqbal, M. U., Nemet, A., & Kravanja, Z. (2019). Recent developments towards enhancing process safety: Inherent safety and cognitive engineering. Computers & Chemical Engineering, 128, 364-383.

Neurosky. (2015). Greek Alphabet Soup – Making Sense of EEG Bands (Last accessed: Nov 10, 2018) http://neurosky.com/2015/05/greek-alphabet-soup-making-sense-of-eeg-bands/

Pfurtscheller, G., Stancak Jr, A., & Neuper, C. (1996). Event-related synchronization (ERS) in the α band—an electrophysiological correlate of cortical idling: a review. International journal of psychophysiology, 24(1-2), 39-46

Liang, Yongqiang, Wei Liang, Jue Qu, and Jie Yang. "Experimental Study on EEG with Different Cognitive Load." In 2018 IEEE International Conference on Systems, Man, and Cybernetics (SMC), pp. 4351-4356. IEEE, 2018.

Morales, J. M., Diaz-Piedra, C., Rieiro, H., Roca-González, J., Romero, S., Catena, A., Fuentes, L.J. & Di Stasi, L. L. (2017). Monitoring driver fatigue using a single-channel electroencephalographic device: A validation study by gaze-based, driving performance, and subjective data. Accident Analysis & Prevention, 109, 62-69.

Sharma, C., Bhavsar, P., Srinivasan, B., & Srinivasan, R. (2016). Eye gaze movement studies of control room operators: A novel approach to improve process safety. Computers & Chemical Engineering, 85, 43-57.

Bhavsar, P., Srinivasan, B., & Srinivasan, R. (2017). Quantifying situation awareness of control room operators using eye-gaze behavior. Computers & Chemical Engineering, 106,191-201.

Iqbal, M. U., Srinivasan, B., & Srinivasan, R. (2019). Electroencephalogram based Biomarkers for Tracking the Cognitive Workload of Operators in Process Industries. In Computer Aided Chemical Engineering (Vol. 46). https://doi.org/10.1016/B978-0-12-818634-3.50233-2

613575 Energy Leadership Toward a Sustainable Energy Future

Abstract:

This presentation will focus on some key leadership principles that will help current and future leaders optimize research toward a sustainable energy future. The presentation will also focus on the current state of fossil fuel usage, and how these fuels will be the bridge toward a future dominated by sustainable energy technologies.

614044 Leadership in Energy Development

Abstract:

Leadership is not simply a trait but a skill essential for organizational success. It is not simply managing but inspiring and creating an organizational will to succeed. Different organizations have different measures of success and require the skill of their people to be uniquely led – and yet there are common elements of leadership that are fundamental in all cases.
Industry, national laboratories, government, academia all have measures of success and a need for strong leadership. I will explore my personal experiences having been in each of these enterprises and will share perspectives for future leaders to consider.

614308 Managing Public Sector Technology Development in a Dynamic Energy Environment

Abstract:

The task of managing energy technology development within a public sector research organization presents special challenges. In addition to the issues faced by private sector managers such as a rapidly changing technology landscape and the ups and downs of the economy, managers of public R&D organizations must also address the need to respond to multiple levels of oversight and direction, changing Administration and Congressional priorities and the constant need to balance short term and long term research needs within the context of uncertain funding levels.

The DOE's National Energy Technology Laboratory maintains a robust portfolio of projects focused on a range of topics that include: (1) improving the level of recovery of oil and natural gas from "unconventional" reservoirs, (2) enhancing oil recovery from conventional reservoirs, (3) reducing methane emissions and (4) accelerating the development of domestic oil and natural gas resources. Managing this portfolio of projects over multiple year timelines requires flexibility, planning for change and special attention to stakeholder communication.

614309 Leading the Nation's Nuclear Cleanup and Providing Sustainability

Abstract:

While most engineers gain acclaim for the enduring projects they design and build, Ken is best known today for what he is demolishing and cleaning up. As President of UCOR, Ken guides and motivates a professional management team and hundreds of skilled workers as they successfully undertake the unprecedented task of removing environmental hazards and millions of square feet of unneeded government-owned facilities from the former Oak Ridge (TN) Gaseous Diffusion Plant – now known as the East Tennessee Technology Park (ETTP).

As a former uranium enrichment site, ETTP presents many cleanup challenges. Buildings planned for demolition are laced with radioactive materials, and years of unregulated waste disposal practices have polluted the soil and groundwater. The extent of contamination was not fully known, and the unsafe, deteriorated condition of many of the structures forced many delays in demolition. Known contaminants include radioactive elements, mercury, asbestos, polychlorinated biphenyls, and industrial wastes.

East Tennessee Technology Park is undergoing a dramatic transformation from its historic national security role as a gaseous diffusion plant – with all the problems posed by aging and contaminated facilities – to a modern private industrial park. As the former gaseous diffusion site is cleaned up, once-contaminated land and buildings are being repurposed for productive use by new job creating private sector ventures. Experiences and lessons learned in the first-ever cleanup of a gaseous diffusion complex anywhere will serve as a global model for reuse of unneeded nuclear facilities elsewhere in the world.

606177 The Challenge for Leaders Today

Abstract:

When women are asked the key characteristics of leadership, they tend to cite confidence and determination, but less than half of them feel that they have these characteristics. The challenge for leaders, then, is to instill these characteristics in others through mentoring relationships, especially to young women in academia. Training the next generation of leaders takes time and personal engagement to teach young colleagues how to plan their career, negotiate and find strong role models. Beyond gender considerations, leaders will find that issues surrounding communication and trust will play critical roles in their effectiveness at promoting an inclusive environment.

602018 Building and Maintaining a Positive Culture in an Academic Department

Abstract:

Developing a departmental culture may sound like the epitome of a fuzzy, non-engineering soft skill, but the long-term differences between organizations with positive and negative cultures can be astonishing. I will reflect on tactics for building a culture that is positive for faculty, staff and students in an academic engineering department. In this environment, it is important to be sensitive to the time pressures faced by people at every level in the organization and the competitive nature of academic research and success.

612621 Introductions of Panelists

Abstract:

Dr. Meredith Sellers is a Senior Managing Engineer in the Materials and Corrosion Engineering Practice at Exponent, an engineering and scientific consulting firm. As a Licensed Professional Chemical Engineer, she assists clients evaluate and investigate failures associated with material properties, processing, and performance. Dr. Sellers has significant experience assisting clients in the oil and natural gas and aerospace industries engaged in domestic and global legal disputes. She holds a B.S. and Ph.D. in Chemical Engineering from Cornell University and the University of Illinois at Urbana-Champaign, respectively. Dr. Sellers is a long-standing member of the Society of Women Engineers (SWE) and the Association for Women in Science (AWIS). She is involved in corporate initiatives within Exponent to both attract and retain women and underrepresented minorities in the workplace.

Dr. Felicia Etzkorn is Professor of Chemistry at Virginia Tech. She has a BS from Missouri State University, a PhD from the University of California, Berkeley, and was an NIH post-doc Harvard Medical School. Her research concerns the stabilization of proteins with small molecules to investigate cell division and collagen structure. She has been the sole PI on 3 major NIH grants, one NSF grant, and multiple smaller grants. She participates in two major NIH biomedical training grants for underrepresented students. She created a Green Chemistry course for senior undergraduates at VT in 2007. Her textbook, *Green Chemistry: Principles and Case Studies* was published in January 2020.

Motunrayo Kemiki is a Renewable Energy and Project Management Professional with more than 10 years experience in the federal government. She currently works at the Office of Renewable Energy Programs at the Bureau of Ocean Energy Management (BOEM) where she is responsible for managing reviews of technical plans submitted for offshore renewable energy projects in the Outer Continental Shelf. Prior to joining BOEM in 2014, she worked as a Project Manager at the Department of Energy Loan Program Office (where she managed technical due diligence of multi-million dollars renewable energy projects -biofuel, energy efficiency & transmission-under Title XVII of the Energy Policy Act) and as a project engineer at the Naval Surface Warfare Center (where she coordinated the manufacturing of rocket motor propellant for the MK6).

She has a Bachelor of Science in Chemical Engineering from the Pennsylvania State University and a Masters in Business Administration from the University of Maryland, College Park. She is also a licensed Project Management Professional.

Named one of the 100 Women Leaders in STEM by STEMconnector, **Tricia Berry** leads efforts to recruit and graduate women in the Cockrell School of Engineering as Director of the Women in Engineering Program at The University of Texas at Austin. She concurrently serves as Collaborative Lead for the Texas Girls Collaborative Project (TxGCP), leading the dissemination of STEM best practices and informal curriculum across Texas in coordination with the National Girls Collaborative Project. Through both roles, she connects and supports organizations and individuals working to advance gender equity in science, technology, engineering and math (STEM) fields across Texas and beyond.

Tricia is the co-author of You Can't Eat Your Degree: Combine Your Passions and Philosophies to Create the Story of Your Future and Exceeds Expectations: Take Control of Your Performance Review. Previously, Berry worked as a Process Engineer and a Product Development Engineer at The

Dow Chemical Company in Freeport, Texas.

Dr. Jessie DeAro is an expert in broadening participation in STEM education research and diversity programs and has nineteen years of experience designing, implementing, and leading federal grant programs at the National Science Foundation (NSF) and the U.S. Department of Education. She has the managed several different diversity related grant programs at the NSF and is currently the NSF ADVANCE program lead and a program director with the EHR Core Research and the Science and Technology Centers programs. She has served in leadership as acting Deputy Division Director at NSF and in policy roles as a senior policy advisor at the White House Office of Science and Technology Policy in 2010.

Ekin O. Gozen is currently the Strategic Portfolio Advisor at Shell located in SF Bay Area. In her current role, as part of Technology Portfolio team, Ekin supports portfolio management and decision making for the Vice President of Refining, Catalyst and Analytical technologies. She delivers cross-portfolio analysis and insights on Shell Technology strategy implementation, productivity of technology portfolio, alignment of portfolio to business technology needs for CTO, the Technology Leadership Team and other key stakeholders. Prior to her current role, Ekin held various roles at Shell Martinez Refinery including the Team Lead for the Quality Assurance group and Process Engineer at Projects team. Ekin joined Shell in 2013 as a Research Engineer in Process Evaluations where she provided technical and economical guidance to various Shell businesses such as Alternative Future Energy, Chemicals, CO_2 management and Upstream. Ekin holds a PhD in Energy Resources Engineering from Stanford University and MSc and BSc in Chemical Engineering from Middle East Technical University in Turkey. In her spare time, Ekin enjoys singing with her band, running and cycling.

601991 The Digital Transformation of the Chemicals Industry - and How to Prepare

Abstract:

Over the course of the next 20 years, the chemicals industry will see more change than it has in the last 50, in large part due to digital technologies. Digital transformation is affecting business processes from R&D, to manufacturing, to supply chain, to sales and go-to-market -- and the data from all these digital efforts is tying those diverse functions to optimize the chemical enterprise and create new business models and sources of value. This presentation looks at how the industry will change through 2040 and examines what the new sources of competitive advantage will be for both companies and individuals.

613114 Driving the Value from Digitalization

Abstract:

Many opportunities exist to utilize the technologies of Industry 4.0 and digitalization within the process industries, to improve reliability, productivity, profitability, and safety. However, organizations may become enamored with the new tool or capability, looking for problems to fit the technology rather than focusing on the true business needs. Only by being technology agnostic and using a strategy that places balance sheet impact and long term competitive advantage as selection criteria to the myriad of solutions available in the market will an organization focus on the proper challenges. This presentation will examine potential pitfalls to technology selection and implementation.

613028 Combining AI with Chemical Engineering First Principles Will Take Process Performance, Safety and Sustainability to New Levels

Abstract:

Digitalization is now in the spotlight in terms of creating improved agility, resilience, and ability to operate process industry assets remotely and safely.

Sensors, connectivity, and cloud computing are making process data more available for use in optimization and prediction. High performance, edge computing, and new algorithms are enabling placing process models online for monitoring, optimization, and operational advice.
Most process and chemical engineers in industry have observed the influx of data science and data scientists into their organizations. This session will cover an area called "hybrid modeling." This combines and embeds AI and machine learning within first-principles rigorous models. You may think of this as putting chemical engineering guardrails around machine learning. Rigorous, first principles models, brings to bear decades of chemical engineering knowledge. Machine learning and AI brings to bear ability to rapidly analyze, and progressively learn from empirical operating data. When combined, it is now possible to achieve a new level of modeling that will further leverage both types of analysis. We have been validating this breakthrough combination of AI and chemical engineering with over 100 companies in industry, and will cover several high value and immediately practical use cases during this session.

601992 Designing Materials and Processes in the Digital World

Abstract:

3M has been known as a materials innovation company for many years with thousands of new products launched annually, many of those enabling and empowering everyday products. This innovation model is based on leveraging deep technologies into novel combinations to develop product platforms.

This talk will address the challenges and progress in transforming this culture and enabling it with digital tools created internally and with outside partners to accelerate the speed of materials innovation dramatically.

AUTHOR INDEX

Bai, Peng....4
Banks, Simon....7
Bardin, Billy....32
Barlaz, David....11
Beck, Ron....33
Berry, Tricia....29
Bright, Zack....18
Cao, Yankai....15
Choudhary, Mahindra....20
Ciferno, Jared....25
Clancy, Paulette....27
Dauenhauer, Paul....3
Dearo, Jessie....29
Dennison, Kathleen....7
Durning, Christopher....12
Etzkorn, Felicia....29
Galownia, Jonathan....7
Gharpure, Akshay....13
Gozen, Ekin....29
Herceg, Eldad....10
Holman, Michael....31
Hsu, Emily....12
Iqbal, Md. Umair....20
Jeckell, Zachary....11
Jurczyk, Brian....11
Kemiki, Motunrayo....29
Khadilkar, Aditi....5
Kieschke, Rob....34
Klara, Scott....23
Krogstad, Daniel....11
Lin, Austin S....8
McConnell, Charles....24
Melocchi, Alice....18
Parietti, Federico....18
Park, Ah-Hyung Alissa....12
Patel, Dhruval....11
Perry, Sarah L....4
Renner, Julie....1
Rueter, Kenneth....26
Ruzic, David....11
Sellers, Meredith....29
Servoss, Shannon....2
Sholl, David S....28
Srinivasan, Babji....20
Srinivasan, Rajagopalan....20
Stefanov, Zdravko....9
Vander Wal, Randy....13
West, Alan....12